The Scottish Coal Industry
Guthrie Hutton

Mines rescue men were the fittest and finest, and went through rigorous training before serving on a team like this.

With somewhat incongruous floral backgrounds, these studio photographs of miners were taken in Motherwell (left) and Dennyloanhead (right).

© Guthrie Hutton, 2022
First published in the United Kingdom, 2022,
by Stenlake Publishing Ltd.
www.stenlake.co.uk
ISBN 978-1-84033-928-4

The publishers regret that they cannot supply
copies of any pictures featured in this book.

Printed by
P2D Books, 1 Newlands Rd, Westoning, Bedford, MK45 5LD

Further Reading

The following were the principal books and websites used by the author during his research. With the exception of *The Wemyss Private Railway* none are available from Stenlake Publishing; please contact your local bookshop, reference library or search for them on the internet.

Brotchie, Alan, *The Wemyss Private Railway*, 1998.
Carvel, J. L., *The Coltness Iron Company*, 1948.
Cunningham, Andrew S., *The Fife Coal Company*, 1922.
Dron, Robert W., The Coalfields of Scotland, 1902.
Duckham, Baron F., *A History of the Scottish Coal Industry*, Vol. 1, 1970.
Industrial Railway Society, *Industrial Locomotives of Scotland*, 1976.
Muir, Augustus, *The Fife Coal Company*, c.1952.
National Coal Board, *Coal* (magazine), Vol. 4, 1950/51.
Oglethorpe, Miles K., *Scottish Collieries*, 2006.
Smith, David L., *The Dalmellington Iron Company*, 1967.

A number of subject specific websites were also consulted, as was Historic Environment Scotland's Canmore site and the National Libraries' NLS Maps site.

Also by Guthrie Hutton, and published by Stenlake Publishing.
Mining, Ayrshire's Lost Industry, 1996.
Lanarkshire's Mining Legacy, 1997.
Mining the Lothians, 1998.
Fife the Mining Kingdom, 1999.
Mining from Kirkintilloch to Clackmannan & Stirling to Slamannan, 2000.
Scotland's Black Diamonds, 2001.
Coal Not Dole, 2005.

Acknowledgements

This little book represents an accumulation of information and material gathered over 25 years of writing about the mining industry. In that time I have met many fine people whose collective knowledge and experience has been invaluable. I have to thank them all. Sadly, a book of this size can offer only a flavour of this once vast industry; I hope the omissions are not too disappointing.

Introduction

Coal was king. It powered factories, heated buildings, propelled ships and drove trains. It was used to smelt iron, make steel, produce gas and generate electricity. It was at the heart of the Industrial Revolution that created wealth to transform the country from a basic rural economy to a modern society.

Before all that happened coal had been mined for centuries from outcrops and small pits, and these were gradually extended to meet increased demand. The big change began after the great Carron Iron Works was started up in 1760. Other works followed as a huge iron and steel industry was established, and that needed vast amounts of coal. Towns and cities with their smoking lums – not for nothing was Edinburgh known as 'Auld Reekie' – were growing and needed more and more fuel. Steam power was developed and that needed fuel, and so through the 19th century the coal industry grew to be very large, very rapidly. Lanarkshire was the main centre, with a huge number of pits working rich seams relatively close to the surface. It was like a gold rush, but coal deposits are finite and by the end of the century the early pits were nearing exhaustion and Fife was becoming the industry's most important area.

After struggling through two world wars and the economic troubles of the inter-war years, the industry was nationalised under a National Coal Board (NCB); vesting day was 1 January 1947. The new arrangement was greeted with optimism, but that faded as other forms of energy like oil and nuclear gained ground. Having invested in huge new collieries, the NCB had to close many old ones, a process that took on its own relentless momentum until only the mines serving the Longannet Power Station were left. They closed in 2002. Opencast workings, that had begun to proliferate in old mining areas, continued to supply fuel for power stations until they too were closed.

No longer king, coal has been cast as the villain by a modern world worried about global warming and climate change, but the history cannot easily be erased. Coal's central role in making the country what it is today is visible in many places. Many villages owe their origins to the mining industry and in some places the environmental impact is hard to miss. Memorials have been erected at various sites, some to commemorate disasters, others proudly to celebrate a once mighty industry. A few pithead installations have been left standing like public artworks. A number of museums tell the mining story, and of these the most significant is the National Mining Museum, Scotland, (formerly the Scottish Mining Museum) in the old Lady Victoria Colliery at Newtongrange, south of Edinburgh. It is a national treasure where the story of coal is kept alive for present and future generations.

The dramatic 180 feet high A-shaped head frame for the No.3 shaft at Barony Colliery in Ayrshire is seen here in the 1950s during construction.

The Scottish coal industry was concentrated across the country's central belt, but there were a couple of outliers including the pit depicted here at Brora in Sutherland. Not only was this, by a considerable distance, the country's most northerly pit, it was also the only one to work coal from the Jurassic rather than the Carboniferous geological period. Initially it was used as fuel for salt pans, but when that became uneconomical the Duke of Sutherland sank the Ross Pit, with a 250 feet deep shaft to the Brora Main Seam to provide fuel for the woollen industry and, later, a brickworks also seen in this picture. When the coal industry was nationalised in 1947 the pit was not taken over by the new National Coal Board (NCB) but left in private hands. Regardless of who owned it, the coal was prone to spontaneous combustion and an underground fire resulted in the pit being sealed in 1969. A subsidiary drift mine remained operational for a few more years, but the colliery finally closed in 1974.

The country's most westerly colliery sat at the southern tip of the Kintyre Peninsula. The early Kilkivan, Shedloch and Trodigal workings, and the Wimbledon Pit, which had been sunk by Hector Ferguson in 1881 and named after a shooting prize he had won there, were all closed by the 1920s. The coal had not been worked out, but it lay untouched until 1944 when the Glasgow Iron & Steel Company proved a seam of up to seventeen feet thick. The NCB took over in 1947 and developed the newly named Argyll Colliery. They drove a couple of drift mines through some very difficult geology; one was steep and intended primarily for ventilation while the other, seen here, was on a more gradual gradient and used to get men and materials in, and coal out. Much of the coal was delivered across the water to power stations in Northern Ireland while some went to the area's perhaps more famous industry, whisky distilling. Spontaneous combustion caused problems, but the shrinking market for coal and difficult geological conditions brought about the mine's closure in 1967.

The village of Rowanburn, in Dumfriesshire, consisted mainly of rows of miners' cottages and a few other buildings like the little school, seen here. The adjacent pit head-frame was part of the Canonbie Colliery, which worked the most southerly coalfield in Scotland. The village took its name from the Rowan Burn, a tributary of the Liddel Water, which at this point forms part of the border with England, placing the colliery only a matter of yards inside Scotland. The coal was worked for much of the 19th century, but activity ceased in 1922. Small and isolated from the main Scottish coalfields, but linked geologically to the West Cumberland coal, Canonbie was an enigma. Mining engineers had for a long time been sure the area had more to offer and so the NCB, with its remit to explore and exploit all of the country's reserves, sunk bores that showed extensive coal deposits concealed under an overlying bed of New Red Sandstone. It was so deep that the industry left it for another day, but the day never came and global warming has probably ensured it never will.

The New Cumnock coalfield in Ayrshire extends into Upper Nithsdale, the northwest corner of Dumfriesshire, where the village of Kirkconnel was transformed in the late 19th century when James Irvine McConnell opened the Fauldhead Colliery. He also opened the smaller Gateside Colliery at Sanquhar. The business remained as a family concern until 1903 when Sanquhar and Kirkconnel Collieries Ltd took over. That company was in turn taken over in 1925 by William Baird and absorbed into the newly-formed company of Bairds and Dalmellington in 1931. The large pits and the Tower Mine, which the Sanquhar and Kirkconnel Collieries Ltd had developed alongside Gateside, were still operational when the NCB took over. They also opened the nearby Roger and Rig drift mines to maximise coal output before major developments elsewhere could be brought into production, but the Roger Mine just kept on going, eventually closing in 1980. Gateside was closed in 1964 and Fauldhead in 1968 despite being in full production.

William Baird & Company started up their Gartsherrie Iron Works at Coatbridge in 1830 and expanded into Ayrshire in 1845, setting up the Eglinton Iron Works at Kilwinning and in the 1850s taking over established ironworks at Lugar and Muirkirk. The iron furnaces needed fuel, so Bairds operated coal mines too, one of which was Highhouse Colliery at Cumnock, opened in 1894. Seen here about 1910 it remained in production until 1983. Although now gone, it will not be forgotten because a head frame has been left standing and an old beam engine, known as Big Ben, which was taken to the Heriot Watt mining school in 1950 has since become an exhibit at the National Mining Museum at Newtongrange. The coalfield extended to the west, so the company began sinking the Barony Colliery in 1906 to exploit it. A new No.3 shaft to a lower depth was begun before the Second World War and completed by the NCB (see page 3). A short time later the original shafts collapsed, a tragic incident that resulted in the deaths of four men. In order to keep the pit working a new No.4 shaft was sunk and completed in 1966. A power station, designed to burn slurry recovered from washery ponds, worked alongside the colliery until 1989 when Barony also closed, the last working colliery in Ayrshire.

When the National Coal Board inherited an industry that had been through two world wars, and a prolonged period of financial and labour troubles, it was in a run-down state and in need of investment. In Scotland, the NCB's 'Plan for Coal' included redeveloping some existing collieries, opening new drift mines and sinking five super-pits, one of which was Killoch in Ayrshire. The area to be exploited lay deep below the surface in the southern Mauchline Basin. It had not previously been touched by the coal industry, but was estimated to contain reserves of over 100 million tons. Work to sink the two concrete-lined shafts began in late December 1952 and by 1957, when this picture was taken, the new colliery was starting to take shape. Killoch was producing coal by 1960 with early results confirming the initial optimism, and in 1965 when the pit became the first in Scotland to produce one million tons a year its future seemed bright, but within a few years geological problems caused output to slow to a fraction of the Scottish average and it never recovered. The pit was closed in 1987 and the impressive surface buildings were subsequently demolished, although the large washing plant was kept functioning to treat coal from opencast workings.

Coal mining was developed in the Doon Valley to provide fuel for the huge new Dalmellington Iron Works, which had been started up in 1848. The driving force behind this was John Houldsworth, whose father, Henry, had established the Coltness Iron Works in Lanarkshire in 1839 and viewed Ayrshire as a logical extension of the business. The new works was located centrally at Waterside with coal workings spread along the valley. Mostly these were drift mines, but there was a large pit, the Houldsworth, near the village of Patna and Pennyvenie Colliery to the south of the valley, close to the town of Dalmellington. It is seen here in a picture showing the No.2 and No.3 pits sunk in 1881. The colliery was eventually worked from a number of pits and mines with the final sinking being No.7 shaft developed by the NCB. Although the Dalmellington Iron Works ceased production in 1921, the associated pits and mines remained in operation and were taken over in 1930 by a new company, Bairds and Dalmellington, which also absorbed Baird's other Ayrshire collieries to become the largest coal company in the county. Pennyvenie closed in 1978.

People tend to think of coal as dirty black stuff and are often surprised to discover that while the industry could do nothing about the blackness, it did endeavour to wash coal before despatching it. Some collieries were equipped with dedicated washers while others had access to a central coal-washing plant such as the one seen in this picture. Situated at Minnivey, it was installed in 1941 by Bairds and Dalmellington, to treat 140 tons of raw coal an hour delivered from the Doon Valley collieries. The contents of the wagons were tipped into a hopper and taken by conveyor to wash boxes and then screens, which graded the coal by size while also separating it from any waste. The slurry and water were treated to ensure that no dirty effluent escaped from the plant. The train passing the coal washer is being hauled by a small steam locomotive known as a 'pug', typical of a type that worked around the country's collieries. It was built in 1953 by Andrew Barclay, Sons & Company of the Caledonia Works in Kilmarnock, one of the foremost makers of these locomotives in Britain. Part of the extensive private railway system that linked all of the company's coal workings in the valley remained operational as an industrial heritage line after the collieries closed.

The most southerly coalfield in Ayrshire was situated in the Girvan Valley, which had a long mining history. Larger workings were developed in the 19th century at Dalquharran, where fumes from an underground fire percolated through the hillside for decades. At the nearby Kilgrammie Pit, a miner named John Brown was trapped underground for 23 days in 1835, but died three days after being rescued. A colliery located at Bargany to the south of the valley, existed in the second half of the 19th century but had ceased operations when a new sinking was started early in the 20th century on the other side of the railway that ran through the valley. Owned by the Killochan Coal Company, it was named Killochan, although locally people still referred to it as Bargany pit. The company went into liquidation in 1914 and was taken over by South Ayrshire Collieries, which also went into liquidation to re-emerge as South Ayrshire Collieries (1928) Ltd. Killochan Colliery and a smaller Maxwell mine were still working when the NCB took over. They opened another Maxwell mine, and a new mine at Dalquharran that remained in operation for ten years after the closure of Killochan in 1967.

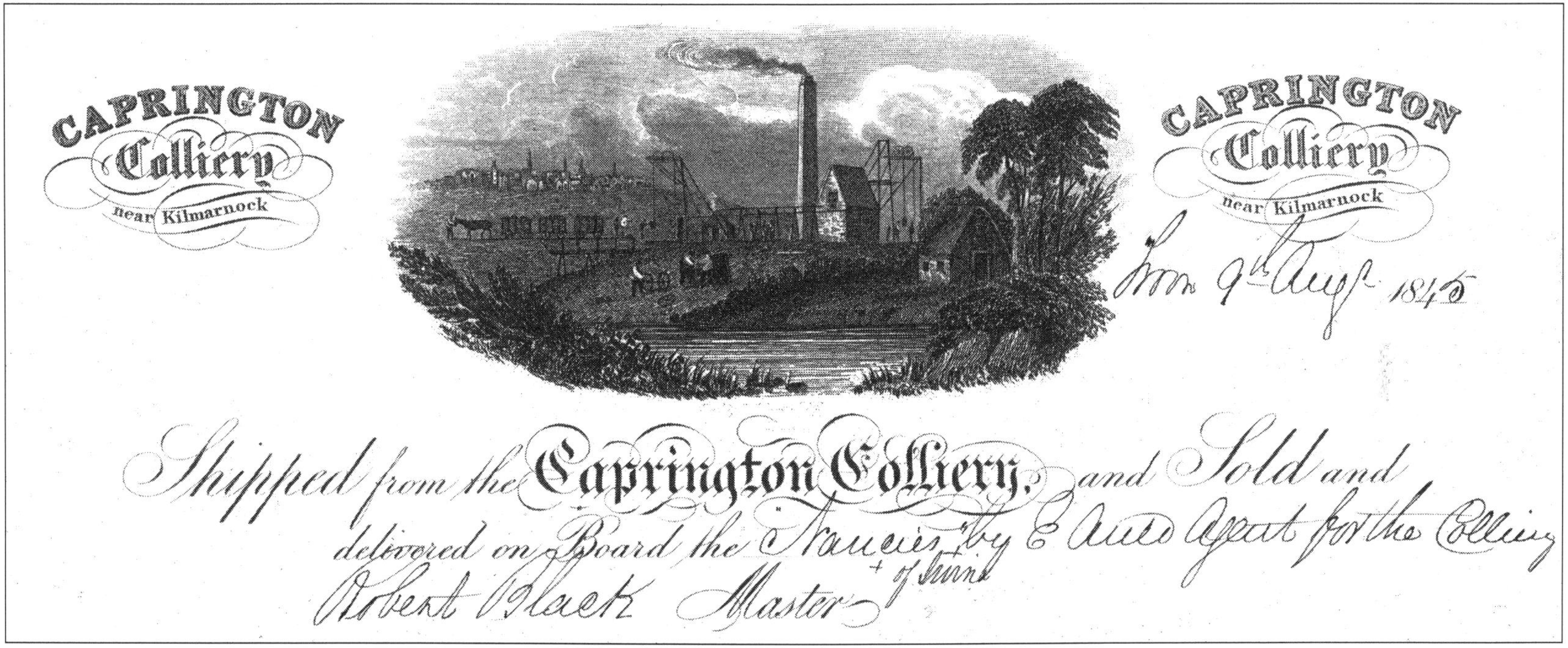

Kilmarnock was an early centre of industrial-scale coal mining. The main instigator was William Bentinck, the Marquis of Titchfield and future Duke of Portland, who, through marriage, acquired coal-rich lands, but his ambitions to exploit this good fortune were hampered by a lack of suitable transport. Ayrshire had no canals, which had helped in coal developments elsewhere, but it was bordered by the sea, the problem was how to get coal to the ships. The answer was to set iron rails on stone blocks between Kilmarnock and Troon Harbour. Wagons were initially horse-hauled, but by 1816 (or 1818) a steam locomotive was in operation, making this Scotland's first railway. Collieries proliferated, with one of the earliest being Caprington. This engraving taken from a bill heading from 1845 may not accurately depict the colliery, but it gives a good impression of an early pit. A more tangible reminder of the colliery is a large beam engine erected in 1811. It was removed to the Dick Institute in Kilmarnock in 1903 and has since become an exhibit at the Museum of Scotland in Edinburgh. The other picture shows a pit at Springside, thought to be Springhill No.3 associated with another of Kilmarnock's major industries, a fireclay works.

Just across the Ayrshire border into Lanarkshire is Coalburn, which although situated in another county was geologically part of the Central Ayrshire coalfield. A stream known as the Coal Burn flowed through the area and its name was applied to the village that grew out of a collection of houses that had built up around the many small pits scattered across the area. After the railway arrived from Motherwell in 1856 the coal industry grew rapidly, with a number of pits that were operated by a succession of owners before Caprington and Auchlochan Collieries Ltd moved in. The pictures show No.6 pit (left), and the large Auchlochan No.9 and No.10 pits which the company developed to the north of the main village in the 1890s. William Dixon & Company, one of the country's major iron and coal owners, took over in the early 1930s and they were still working the Auchlochan pits in 1947 when the NCB moved in. They managed the pits as part of their Ayrshire Area, but closed them all by 1968.

One of the small early collieries near Coalburn was named Dalquhandy. It was also the name given to a huge opencast coal site developed close to the village some 20 years after the last of the Auchlochan pits closed, seen here in 1999. It was one of many such sites opened in former coal mining areas across Scotland. They employed fewer men to produce more coal per man shift than deep mines ever could, but had a major impact on local communities and the environment. Huge machines dug into the ground over vast areas to win coal the miners working underground had left behind. Often this was from an early method of working known as 'stoop and room' in which large blocks of coal were left in place to support the roof, but also from seams untouched by deep mining. The coal was moved off site by very large dump trucks and then taken to a washing plant before despatch to power stations. When fossil fuel was being dug at Dalquhandy it was described as the largest hole in Western Europe. Ironically since then trees have been planted on the former site to absorb carbon and wind turbines erected to generate renewable energy.

At Larkhall, to the north of Coalburn, there were a number of pits including the Bog Colliery, seen here. It was sunk on the Duke of Hamilton's land and operated by Hamilton, McCulloch & Company to work the Ell, Upper and Lower Drumgray, Pyotshaw, Main, Splint, Kiltongue and Virtuewell seams, all familiar names to mining men in this part of Lanarkshire. An ell was an old Scots unit of measure, about the length of a man's arm, but in the Larkhall area it was misnamed being frequently found at between ten and twelve feet thick. A soft fireclay often overlaid the seam, which meant that up to four feet of coal had to be left intact to support the roof. The company also worked the nearby Home Farm Pit. It was the scene of a disaster in 1877 when water, mud and sand burst out of the Ell coal workings into one of the shafts. Most of the 60 men escaped the flood, but four died. Three men repairing a shaft at the Bog Colliery were killed in 1894, an explosion there in 1889 injured seven and many other accidents at these and neighbouring pits resulted in single fatalities or injury. The Bog Colliery closed in 1927.

John Watson's father, also John, was a stonemason from Cupar in Fife who moved to Kirkintilloch in 1814. He enjoyed success as a builder, but was soon attracted to the coal trade. Young John also got involved, working at his father's colliery and on the canal barges taking coal to Glasgow through the night. It was a tough apprenticeship, but John Watson (Jnr) was soon branching out on his own account. He sunk pits at Wishaw, Slamannan and Motherwell before turning to the Hamilton area where he bought the Neilsland and Earnock Estates in the 1870s. His new Earnock Colliery, which started to win coal in 1879, is seen here in a picture from about 1910. It was a model colliery with the shafts arranged to facilitate escape in the event of an explosion – an uncomfortably common experience in this part of Lanarkshire. The pit was the first in Britain to have electric lighting installed at the shaft bottoms and principal roadways, and was also used to pioneer telephone communication between the surface and pit bottom. A new shaft was sunk to the Splint coal in 1891 and for a time Earnock employed more miners than any other Lanarkshire pit. It closed in 1942.

With its centre around Hamilton, Uddingston and Cambuslang, the Lanarkshire coalfield contained many thick and valuable seams at workable depths, but historically was only exploited on a modest scale. Then, as the market and railways expanded, the pace quickened until in the later 19th century and into the 20th Lanarkshire came to account for half of the Scottish coal industry, with more pits, more miners and greater output than the rest put together. There were a very large number of small or medium sized pits, like the one in this picture, which is believed to have been in the Hamilton area. One reason for this was the get-rich-quick tendency of landowners to issue small leaseholds, making for intensive development but leaving little for the future. It also sowed the seeds of demise because, by the time the NCB took over, the field was nearing exhaustion and large areas of old workings were so completely flooded that dewatering and redevelopment were impractical. Consequently, the NCB chose to site their big developments in Fife and the Lothians and encouraged Lanarkshire mining families to move east to the new pits.

One typically small Lanarkshire colliery was the Farme Pit at Rutherglen, which although it looked less than impressive on the surface, had a long and productive life. So too did James Anderson, the pit manager who, when he celebrated 50 years in the job in 1911, claimed that all the roadways made in his time 'would, if joined together, make over 35 tunnels from England to France and be equal to a main road of over 1,000 miles'. Like so many other pits in the county, Farme would probably have disappeared without trace had it not been for a remarkable piece of machinery, an atmospheric condensing steam engine. Known at the pit as the 'old engine', it had been installed in 1809 when the Old Farme No.1 Pit was sunk and during its working life raised about three million tons of coal to the surface, each lift taking about 35 seconds. It is seen in this picture in front of the wooden head-frame. The dome-like object on the right was not part of the engine. Farme, which was operated throughout its existence by a private company, closed about 1916.

Lanarkshire was Scotland's industrial heartland, replete with iron furnaces, foundries, steelworks and heavy engineering workshops that all needed vast quantities of coal. The county's great advantage was that its Splint coal could be fed directly into iron smelting furnaces without the need to be coked (coal baked to remove gases) as was required by furnaces elsewhere. Unsurprisingly, the owners of the great blast furnace plants also operated collieries to guarantee fuel supplies. One such was the Coltness Iron Company's Milnwood Colliery at Bellshill, seen here in 1901 about five years after its opening. A power plant provided electricity to the underground haulage system, pumps and lighting, and also electric lighting in workers' houses. An improved method of screening, designed by the general mining manager for the company's sixteen pits, was installed on the surface. Large coal from Milnwood went to the blast furnaces while the small coal was washed and separated for sale into three sizes. The pit was latterly owned by the Wilsons and Clyde Coal Company, which had close links to the Coltness Iron Company, and closed before Nationalisation.

Another iron industry owner, William Baird & Company opened Bedlay Colliery at Annathill, in the north of the county, in 1905. The deepest of its three shafts went down to over 1200 feet to reach thin seams of high quality coking coal. Despite being very gassy it was regarded as a good pit to work in, perhaps exemplified by the attitude of the miner who went to work when the pit was nationalised dressed in a suit, shirt and tie, like any other civil servant. Miners always did have a keen sense of humour. In the early 1950s when the decision to build the Ravenscraig Steel Works at Motherwell was taken, the NCB implemented a plan to double the pit's output. The headgear for the three shafts was upgraded, No.1 was renewed, the head frame from No.3 shaft was moved to No.2 shaft and a new concrete winding tower was erected over No.3 shaft. It was equipped with a four rope electric friction winder and double-decked cages that could accommodate 48 men. Latterly a place of pilgrimage for pug locomotive enthusiasts, Bedlay was closed in 1981.

Through the Kelvin Valley, to the north east of Glasgow, there were a number of pits including Dumbreck, with its coke ovens and fiery flares. Twechar Colliery Jwas situated on both sides of the Forth & Clyde Canal while Wester Gartshore and Gartshore sat beside the Edinburgh and Glasgow Railway. The line is seen in this picture from 1912 with a train that has been stopped to allow a boiler to be unloaded – not the kind of disruption modern railway operators, with their fast services and tight timetables, would welcome. The boiler was intended for the Dullatur Pit opened by Baton Collieries in 1913, but closed after only a year at the start of the First World War. A new Dullatur mine, opened by the Cadzow Coal Company, went into production in 1935. It worked under the Dullatur Bog, which made the mine a very wet one to work in. When the NCB took over they pushed the workings under the Forth & Clyde Canal, designing them to ensure that any subsidence did not migrate to the surface and damage the integrity of the canal. Dullatur closed in 1964, the bog was declared a Site of Special Scientific Interest, and the canal reopened in 2001.

The Dullatur Bog was the Central Scotland watershed sending the River Kelvin to the west and the Bonny Water east toward the Firth of Forth. At the head of the Bonny Valley is Banknock where the colliery depicted on this old picture postcard was located. The quality of the printing is poor and the image indistinct, but the pit head-frame can be seen on the extreme left with the waste bing to the right. The postcard is captioned 'Removing Banknock Jewels', perhaps a reference to the railway locomotive on the left, which could be hauling away wagons loaded with coal from one of the area's principal seams, the Banknock Jewel. The Banknock Coal Company operated local pits and there were more collieries further down the valley worked by other owners, but the closing of pits in the 1920s continued until Robert Addie's Herbertshire Colliery at Denny was the only one remaining in the Bonny Valley at Nationalisation. It closed in 1959. A number of small private mines also worked in the area around Banknock and Longcroft.

Tucked in behind a wall in this picture from the eastern edge of Stirling, is a colliery with two head-frames, two bings and two names: Millhall and Polmaise No.1 and No.2. Archibald Russell, who started mining coal at Cambuslang but originally hailed from Clackmannanshire, sunk the pit between 1902 and 1904. Two years later he opened another pit, Polmaise No.3 and No.4 close to the southern shore of the River Forth. There was no convenient housing, as there had been in Stirling, so the company had to create the new village of Fallin for the miners and their families. The two pits worked the same seams, with one of the most valued being anthracite (see inside front cover), a particularly pure coal that burns at a fierce heat without smoke. Found in only a few places, it was highly sought after – Welsh anthracite was famous – and this corner of Stirlingshire was a major producer with Manor Powis, another large pit on the north shore of the Forth, also mining high quality anthracite. Despite starting at much the same time, the fortunes of Millhall and Fallin diverged with the former closing in 1958, while the latter, better known as just Polmaise, was redeveloped by the NCB before eventual closure in 1987.

Like Fallin, Cowie was another new pit village to the southeast of Stirling. It was built by the Alloa Coal Company to accommodate miners and their families employed at the Bannockburn Colliery, which they opened in 1893/95 to work high quality coking coal. Legislation authorising local authorities to build houses was first enacted in 1909, but had little impact until further Parliamentary acts after the First World War began a concentrated period of council house building. Prior to that, large industrial enterprises in hitherto unpopulated places had to provide housing for the workers. Usually this took the form of small cottages built in rows and Cowie was no exception. It wasn't pretty, but it was sufficiently large to warrant a school, church and recreational facilities. It also had a branch of the Bannockburn Co-operative Society as seen in this picture. Formed in 1830, the society was one of the oldest, predating by two years the Rochdale Pioneers regarded as the model on which the Co-operative movement was built. The Co-operative 'store' was a popular feature in many mining villages.

When miners' leader Alexander McDonald became an MP in 1874 he immediately drew attention to the appalling housing conditions endured by many mining families. Sensing a story, a journalist from the *Glasgow Herald* went looking and wrote up his findings in a series of articles headed *Notes on Miners' Houses* that collectively form a remarkable social history document. He didn't sensationalise, he didn't need to, the dispassionate descriptions of what he found were devastating. Common problems included poor building construction and maintenance, cramped living spaces and foul, insanitary arrangements for dealing with household and human waste. Dampness was everywhere. In one area he found that; 'the houses are all bad and for the most part uninhabitable in their present condition. In one . . . the wall paper of the room, owing to the damp, was peeling off in strips, the bedclothes were moist from the same cause, the furniture was getting out of joint, and indeed the entire contents of the apartment were falling into decay.' For one group of houses, the reporter concluded that 'the only way to improve them is to sweep them away'. This miners' row at Low Quarter, near Hamilton, was typical.

Recreational activities developed widely in mining communities, and of these one of the most popular was music, played by either brass or pipe bands. Provision of uniforms and instruments could be expensive so deductions were made from pay packets to provide a fund. A fully kitted-up pipe band, like this from the Michael Colliery in Fife, could be especially costly, but the ability of bands to win prizes brought pride to the local community. Football teams also won prizes and gave local communities a sense of identity. Some had colourful names like Crossgates Primrose, Benquhat Heatherbell and Newtongrange Star although the most remarkable has to be the Ayrshire team, the Glenbuck Cherrypickers. Men from mining areas were fiercely competitive and many graduated from their local leagues to achieve high honours at club and international level. Some preferred more gentle pursuits like pigeon racing 'fleein doos', while others liked nothing better than tending to their allotment gardens, or performing in drama competitions. Amateur groups around the country regularly performed the plays written by Cardenden miner, Joe Corrie.

In an effort to redress some of the social difficulties in pit communities, the Miners' Welfare Fund was established in 1920. A levy of a penny on every ton of coal, plus contributions deducted from miners' pay packets was used to build up the fund. The money had to be spent in the areas where it was raised, and local committees made up of management and workforce representatives were formed to determine priorities. Most areas chose to build institutes, which provided a focus for social life, with facilities like a recreation hall, games room and a library. Some institutes also had baths to compensate for the lack of such facilities in houses. Outdoor facilities also formed a significant element at some institutes. Typically these might include gardens, tennis courts, children's play park and, almost always on the list, a bowling green. The institute shown here at Burngreen, Kilsyth, had all four. The welfare provision continued after Nationalisation under the auspices of a new organisation, the Coal Industry Social Welfare Organisation, CISWO.

The Miners' Welfare Fund also allowed for the provision of convalescent homes. These were favoured in Ayrshire where a large mansion, Kirkmichael House was fitted up to provide facilities for men and boys. Another property was acquired at Troon in 1924 and made available for women and girls. Known as Portland Villa, it is seen here in 1938, a couple of years after being extended with views across the gardens to the sea. It became a convalescent home for men and women following closure of the Kirkmichael Home in 1956. The Ayrshire coast was evidently a favoured location because a convalescent home for Lanarkshire miners was also established at Saltcoats. In Fife, the C. Augustus Carlow Memorial Home for women and girls was opened at Leven in 1948. Prior to that, Blair Castle, a large mansion near Culross, became Fife Coal Company property when they bought the estate in 1927 for mineral rights, and then gifted the house to the Fife and Clackmannan Welfare Committee for use as a male convalescent home. Named the Carlow Home, it became the single convalescent home in Scotland for men and women as the industry contracted and other homes were closed.

One grim reality for the mining industry was the high number of incidents and accidents resulting in injury or death. Often only one or two men were involved, but some were major disasters with multiple casualties. As pits went deeper, primitive ventilation systems struggled to cope allowing a build-up of methane gas, also known as firedamp, which when combined with air made for a dangerous mixture. Eighteen men were killed in an explosion at Commonhead in Airdrie in 1850 and the following year 61 died in another explosion at the Victoria Pit at Nitshill, at the time Scotland's deepest pit. The worst disaster in a Scottish pit occurred in 1877 when 207, and possibly more, men died in a series of explosions at Blantyre. It was such a shocking event that it was featured in periodicals like the one shown here. Steps were taken to hopefully avoid any repetition, but 27 men died in another explosion at Blantyre and 73 were killed at Udston Colliery, Hamilton, in 1887. And there were more fatalities from explosions when two men were killed at Denny in 1883 and thirteen at Dunipace in 1895. For some people, the price of coal was high.

Explosions were not the only cause of disasters. Fires, floods, falls of ground and other accidents proved equally fatal. In an attempt to improve safety, the Coal Mines Act of 1911 required Coalowners Associations to set up Mines Rescue Stations and to train men in rescue techniques. The Fife & Clackmannan Association established its station at Cowdenbeath in 1910 and other areas followed with stations at Kilmarnock, Edinburgh and Coatbridge. The finest hour for the Mines Rescue Service came in 1950 when a heading being driven at Knockshinnoch Castle Colliery in Ayrshire broke through the strata into a basin of peat, which flooded into the pit trapping 129 men. The only way out was through old gas-filled workings of the neighbouring Bank Colliery and the only way for the men to negotiate these was with breathing apparatus that none of them had previously used. The Mines Rescue men broke through between the two pits to make a connection and brought out 116 fellow miners in relays of three men at a time. It was a triumph tempered by the knowledge that thirteen men, who could not be reached, died. Rescuers and rescued are seen in this dramatic picture.

Labour relations in the mining industry were often fraught. Initially men were employed at a single pit and had no bargaining power, but gradually a union structure developed. County unions were formed and these grouped together in 1894 as the Scottish Miners' Federation, which affiliated to the Miners' Federation of Great Britain (MFGB). A strike that year collapsed, union membership fell and, for a few years, pay fluctuated along with coal prices. In 1909 the owners proposed a pay cut coupled with a lockout threat, which led to a tense stand off that dragged on until 1911 when the MFGB demanded a fair minimum wage. Government intervention failed to resolve the situation and in March 1912 the first national miners' strike began. People in mining communities soon began to struggle financially and set up soup kitchens to provide food. Others dug for coal in colliery waste bings or old workings. Bands like this one from Blackburn West Lothian were formed and went around the country performing to raise funds. The strike lasted six weeks and gained little but the miners had learned that when they acted together owners and government listened.

During the First World War the government controlled coal prices, but abandoned these measures when the war ended. The price of coal dropped and unable to meet high costs on low returns the owners proposed a pay cut. The miners rejected the idea and went on strike in 1921. It was a bitter dispute and the armed forces, so recently fighting a war, were deployed to guard pits, as this group in the so-called Sergeant's Mess at Leven Pit in Fife have done. The dispute lasted for three months after which time the miners went back to lower wages and fewer jobs. Trouble broke out again in 1926 when the miners went on strike and other trade unionists joined them in the General Strike. It lasted for nine days but the miners held out for seven months. Nothing was gained and jobs disappeared as more pits closed. When the industry was nationalised after the Second World War the National Union of Mineworkers (NUM) was formed at the same time. Hopes for a rejuvenated industry were high, but peace proved illusory. Frequent small disputes in the 1960s were followed by strikes and an overtime ban in the 1970s and the last big dispute, the tumultuous miners' strike of 1984/85.

Underlying geology didn't neatly conform to later county boundaries, so the coal worked in south western West Lothian was the same as that in Lanarkshire. There were a number of collieries in the area prior to Nationalisation as this picture looking east from Westcraigs toward Armadale shows. The large bing in the centre of the picture was formed by waste from Westrigg Colliery, which was never a high performer, eventually closing in 1930. In common with other local pits, Westrigg also produced fireclay, which was used to make bricks in the pit's own kilns or taken to the Bathville Fireclay Works at Armadale. The pit was latterly operated by United Collieries Ltd., a grouping of small coal companies many of which were situated along the Airdrie – Bathgate railway, the one in this picture. The railway yards at Bathgate were a hub for many local pits including Easton Colliery, one of the largest and most successful in West Lothian. Further south the NCB inherited a number of pits including the big Polkemmet Colliery, at Whitburn. It latterly produced vital coking coal for Ravenscraig Steel Works, but closed after being flooded during the 1984/85 Miners' Strike.

Situated at Bo'ness on the southern shore of the Firth of Forth, and separated geologically from the other West Lothian pits, the large Kinneil Colliery was the latest in a long line of local coal workings. The industry was given a huge boost when the great Carron Iron Works started up nearby in 1760. One of its founders, Dr John Roebuck, invited a young James Watt to develop a steam-driven pumping engine for a flood-prone pit, but when Roebuck was declared bankrupt Watt had to leave Bo'ness to continue his work in Birmingham, in partnership with Matthew Boulton. Almost two centuries later, when the NCB took over at Kinneil, the pit had been struggling for a while, but just tapped a new seam of good coking coal. Further investigation prompted the Board to redevelop the pit with two new shafts. The tower and associated car hall for No.1 shaft are seen here under construction in 1956 with No.2 shaft on the right. Predictions of the pit's likely productivity proved wide of the mark, but in 1964 a tunnel driven under the Forth Between Kinneil and Valleyfield Colliery in Fife boosted the two pits, with the output of both being treated at Kinneil. Fife can be seen in the background, across the Forth.

As Lanarkshire's once mighty industry declined, Fife became the country's leading area. It's not that coal had just been discovered, the Abbot of Dunfermline had been given permission to extract it from Pittencrieff Glen in the 13th century, but the ability to work deep rich seams on an industrial scale allowed the industry to develop. To the north of Valleyfield, the Carron Iron Company opened Blairhall Colliery in 1870 primarily as an ironstone pit. They sold out after a short time to the Lochgelly Iron Company who, in turn, sold it to one of the giants of the iron industry, the Coltness Iron Company – emphasising the importance of coal to that industry. They worked the pit for a number of years before sinking two new shafts to reach the Jersey, Mynheer, Glassee, Five Feet and Dunfermline Splint seams. When the NCB inherited the pit, they chose to upgrade it to increase production by installing underground locomotive haulage and carrying out surface reconstruction including a new coal preparation plant, seen on the right of this picture taken in 1956 while the works were in progress. Blairhall closed in 1969.

There was a large concentration of pits in West Fife. Some were situated around the nation's ancient capital, Dunfermline, but King Coal's capital was Cowdenbeath. Pits were all around. The Cowdenbeath Coal Company provided the early impetus but development quickened when it was taken over by the Fife Coal Company in 1896. Cowdenbeath became a boom town, with coal at its heart. The population doubled. The Fife Mining School was established in the town, so too the mines rescue station for Fife and Clackmannan and a large miners' welfare institute with bowling green and tennis courts. Known as 'The Miners' (and later the Blue Brazil, but that's another story), Cowdenbeath Football Club became a force to be reckoned with. Their home ground, Central Park, was built up with colliery waste and had a fine view of No.7 Pit. The Fife Coal Company also established a large central workshop and offices in the town to service the needs of all of their pits. It superseded earlier workshops that the men in this picture were evidently proud to promote. The NCB continued to use the central workshops and offices for the Fife and Clackmannan Area and despite the closure of local pits they remained in use until 1988.

Mining had been carried on in the district around Kelty for many years before the Fife Coal Company moved in and set about proving the deeper seams. Encouraged by the result, they began sinking two new shafts in 1874. Although numbered officially as Kelty No. 4 and No.5, the shafts were in reality a new pit that was named the Lindsay Colliery after the company's first chairman William Lindsay. Not content with one new pit, the company sank another shaft a couple of miles to the north, linked it underground to the Lindsay and named it after Thomas Aitken who had succeeded William Lindsay as company chairman. The Aitken Colliery proved to be a consistently strong performer, an occasional record breaker, and a jewel in the Fife Coal Company's crown. The company also sited a large power station at the Aitken Colliery to provide electricity to all of its pits in the area, an arrangement continued by the NCB after they took over. The two pits remained in production under the NCB, but not for long, The Aitken closed in 1963 and the Lindsay, known affectionately as the 'Grand Old Lady' and seen here in a picture from about 1910, followed in 1965.

Lumphinnans, Lochgelly, Glencraig, Lochore and Ballingry formed a string of mining villages to the south and east of where lovely Loch Ore had been before the landowner drained it in 1798. Some of the associated pits took the names of the villages; others were named Nellie, Jenny Gray or Mary. The Fife Coal Company and Lochgelly Iron and Coal Company were the main operators, with the Wilson's and Clyde Coal Company, an associate of the Coltness Iron Company, working Glencraig Colliery seen here about 1910. Sinking of the two rectangular shafts began in 1895, but one was distorted by subsidence leaving the pit with a history of shaft accidents. Taken over by the NCB it remained in production until 1966, a date that roughly coincided with the closure of the neighbouring pits. They left a legacy of a blighted landscape and bings that burned at fierce temperatures polluting the air. To sort out the mess, Fife County Council embarked on a vast land reclamation project. The bings were levelled, old structures demolished, the area landscaped and the award-winning Lochore Meadows Country Park created. One reminder of the industry remained, the concrete head frame from the No.2 shaft of the Fife Coal Company's Mary Colliery.

There was another cluster of pits associated with the villages of Auchterderran, Bowhill, Cardenden and Dundonald – Fife's ABCD. Of these pits, the largest was Bowhill, sunk in the late 1890s by the Bowhill Coal Company and taken over in 1909 by the Fife Coal Company; their wagons are prominent in this picture. One of mining's ever-present dangers afflicted the pit in 1931 when ten men were killed in a firedamp (methane) explosion as they were carrying out adjustments to the ventilation system. Just prior to Nationalisation the company initiated a reconstruction scheme to improve the efficiency of conveyor systems and haulages, along with the introduction of locomotives, mine cars and mechanical handling devices for the cars. Surface reorganisation was also planned. The NCB continued with these developments and went much further, adding a new No.3 shaft to the scheme. It was to be sunk to 2760 feet with a level mine driven from its base and another at an intermediate point to intersect with the dipping seams. A new concrete head frame rose above it and was operational by the early 1960s, but the optimistic output targets were never met and the pit closed in 1965.

If Bowhill's early demise was a disappointment, a few miles to the east at Rothes Colliery the NCB had to confront a much bigger embarrassment. The pit was planned by the Fife Coal Company, which held the ceremony to dig the first sod in December 1946, before Nationalisation. Actual sinking began in 1948, but soon water-bearing sandstone was encountered and as the two shafts inched downwards problems with ground conditions mounted. The normally wet Kinglassie Colliery nearby, became hot and dry; this was not going well. A level horizon was opened out at 1600 feet from No.2 shaft and the pit raised its first coal from the Five Feet Seam in 1957, just in time for the Queen to don a white boiler suit and visit the colliery in 1958. But results from the upper levels were even poorer than expected and attempts to drive dipping mines to lower levels were stopped in 1962. The showpiece colliery had proved to be an expensive failure. The NCB chose to get all the bad news out at the same time by also announcing the closure of another big failure, Glenochil Mine in Clackmannanshire. The nationalised industry had lost its shine.

When King James VI described Fife as 'a beggar's mantle fringed with gold' he could equally have substituted black diamonds for gold because coastal collieries working undersea seams produced great riches. At Nationalisation the top two pits in the country were the Michael Colliery at East Wemyss and the Wellesley Colliey at Methil. They were both worked by the Wemyss Coal Company, which also operated other coastal pits from the Victoria and Lochhead in the south to the Rosie and Muiredge at Denbeath. They were all linked by the Wemyss Private Railway, which with 45 miles of track was Scotland's largest private railway, and even after the NCB took over the colliery sidings, the fifteen mile long main line remained in private hands. A pug locomotive hauling an NCB wagon is seen here with the Wellesley Colliery in the background. Begun by Bowman and Company as Denbeath Colliery in 1883 it was taken over by the Wemyss Coal Company in 1905, and subsequently renamed. Its most distinguishing feature was a huge 'Baum' washer, which treated coal brought from all of the company's pits by the private railway. Wellesley closed in 1967.

Much of the output from the coastal collieries was destined for export and much of the shipping was done through the port of Methil. The first dock was opened in 1887, a second in 1900 and the large No.3 Dock in 1913. Coal hoists, like the one shown here, were used to elevate railway wagons and tip their contents down a chute into ships' holds. The dust cloud indicates that this was not the cleanest way to handle coal, but it was quick, and in the port's heyday there was always another ship waiting to be loaded. The full potential of No.3 dock was never achieved. Coal industry output peaked in 1913, the First World War broke out the following year and post war difficulties reduced exports to European destinations. Despite all that, output from the Wemyss pits remained high, but in common with much of the industry in the 1960s the coastal pits began to close. Following closure of the Wellesley Colliery, the Michael Colliery shut in 1967 after a dreadful fire in which nine men died. Further south, the Frances Colliery at Dysart closed after the strike of 1984/85 and another of the great NCB developments, Seafield Colliery at Kirkcaldy, which opened in 1966, closed in 1988.

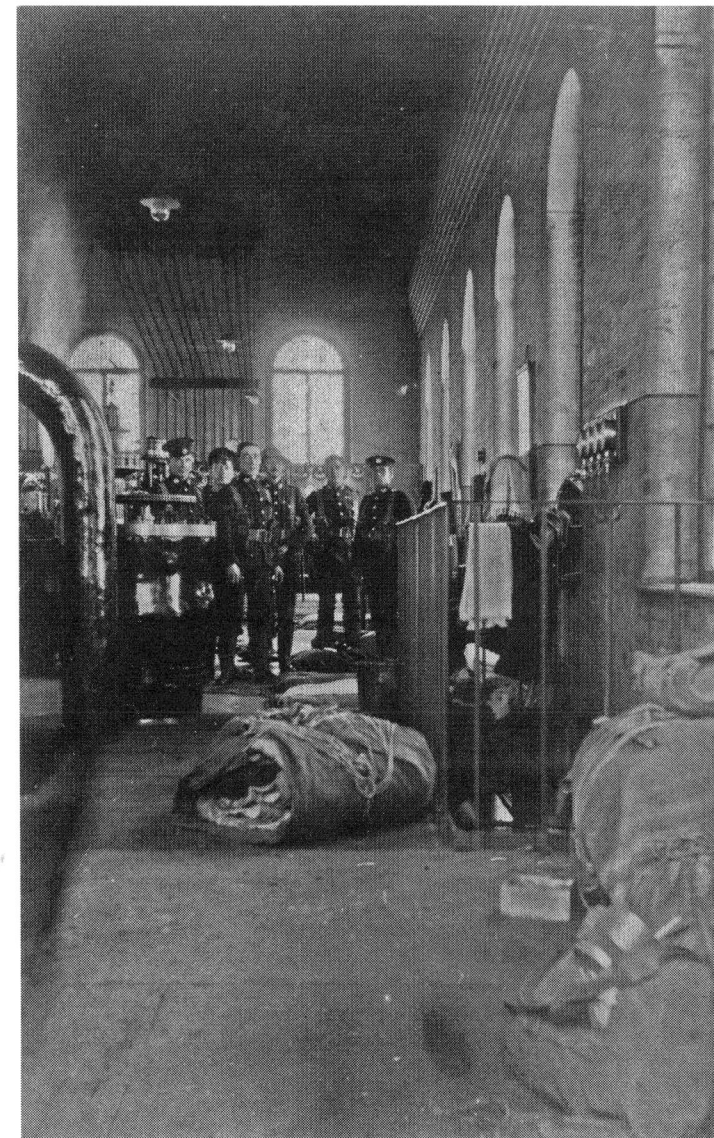

Salt pans were located at a number of places along the Fife and Lothian shores of the Firth of Forth. The pans were large shallow iron troughs raised off the ground so that, when filled with sea water, they could be heated by burning small coal known as 'pan wood' under them. It was a lucrative business until ended by tax changes and cheap imports, but some place names like Prestonpans remain as a reminder. Coal for the salt pans came from outcropping seams close to the surface, but there were other, deeper seams that the Grant-Suttie family of Prestongrange House worked from time-to-time. In 1895 a company, that later became the Summerlee Iron Company, took over, deepened and redeveloped the pit. As well as coal, Prestongrange Colliery produced a good quality fireclay and so a brick and tile works was sited alongside. It outlasted the pit, which closed in 1962. That was not the end, because a large beam engine that had been installed in 1874 and taken out of service in 1954 was still there. Too good to demolish it formed the nucleus of a Scottish Mining Museum and remains as an industrial museum exhibit. It is seen here alongside a picture taken in the pit's other engine house by troops guarding the colliery during the 1921 strike.

Carberry Colliery, seen here in the 1920s, was situated near the historic Carberry Tower in a landscape traversed by the principal approach road to Edinburgh from the east where two battles were fought, Pinkie in 1547 and Carberry Hill 20 years later. Opened in the 1860s to mine the Great Seam, the pit was originally worked by coal owners Deans and Moore who also operated the nearby Wallyford Colliery and other pits and mines in East Lothian. It was taken over in 1900 by The Edinburgh Collieries Limited. They sank a new No.3 shaft to open up other seams and in the process made the existing No.1 shaft redundant. The original shafts had been sunk to 450 feet, but as the workings progressed two faults had to be overcome, each of which dropped the coal measures by about 150 feet taking the depth to 750 feet. When the NCB took over they estimated a long life for the pit, but closed it in 1960. The Edinburgh Collieries had also predicted a long life for Wallyford Colliery when they took over there, but it was closed before Nationalisation.

The Benhar Coal Company originated at Benhar near Harthill in Lanarkshire, close to the border with West Lothian. It began moving its interests to the east in 1882 when it took over the lease on the Niddrie Colliery, an operation characterised by numerous pits and mines. The business also changed its name to the Niddrie and Benhar Coal Company and in 1897 began sinking a new colliery beside the village of Newcraighall. Although the area had been heavily worked in the past, the new shaft was intended to gain access to an enlarged area and go deeper, to 800 feet, with workings extending to the north beyond the Musselburgh shoreline. Perhaps because the new pit was predicted to tap untold riches for over a hundred years it became known as the 'Klondyke'. The company also enlarged the village of Newcraighall, which was absorbed into the City of Edinburgh as a result of boundary changes in 1920. It is seen in this picture from 1930 looking west along Whitehill Street with the colliery in the distance. The village survived the closure of the colliery in 1968 and the area has since become a destination, with a new railway station, park and ride facility and nearby retail park, Fort Kinnaird.

Engineers and miners had to contend with a variety of geological conditions, but few were as challenging as those encountered at Gilmerton Colliery, just south of Edinburgh. There the strata was so distorted the seams lay almost vertical requiring a totally different mining technique. Entering a seam by way of dooks from a main haulage road and, with the roof and floor on either side of them, the miners worked their way up and down an area of coal. They operated in pairs with one man winning the coal and the other loading. Not all miners were comfortable working in 'the steeps'. It was dangerous; the pick man could be operating many feet above his colleague who was protected from falling coal only by a scaffold made up of pit props, or 'timbering' as it was known. The miners ensured that these props were very secure as they also acted as access ladder and working platform. A good indication of what this topsy-turvy way of mining was like is conveyed by these illustrations, drawn for the NCB magazine *Coal*, in July 1950. The near vertical seams proved to be the pit's downfall in 1961 when they acted like a chimney for a fire burning deep underground. The pit was sealed off and closed.

Despite vociferous protests, the railway known as the 'Waverley Line' that ran from Edinburgh through the Border country to Carlisle was closed in 1969. Partially reopened since then as the Borders Railway, the line also served a number of pits including the Arniston Colliery at Gorebridge, which is seen here with the railway in the foreground. The Gore Pit, in the picture, was sunk by the Arniston Coal Company in 1874-78 and worked in conjunction with the older Emily Pit until 1962 when the colliery closed. To the north of Gorbridge, at Newtongrange, the railway served the Newbattle group of collieries. Coal had been mined in the area for centuries before the Lothian Coal Company began sinking the Lady Victoria Colliery in 1890 and linked it to the older engine pit renamed as Lingerwood. The group's third unit, Easthouses Mine, was opened in 1909. One of the industry's biggest and best pits, the Lady Victoria remained working until 1981. As a near-perfect surviving example of a large Victorian colliery, it took on a new life after closure as the Scottish Mining Museum, later restyled as the National Mining Museum, Scotland, a splendid celebration of a once-proud industry. When the railway line was reopened a walkway was created to link the new Newtongrange Station to the museum.